Grow the Best Peppers

Weldon Burge

CONTENTS

Introduction

For American gardeners, peppers are second only to tomatoes in popularity. They produce well in limited space, are virtually free of pests and diseases, and are fairly easy to grow. The plants are attractive in the garden; many people grow several varieties just for their decorative touch.

Bell peppers are the most popular and familiar, but in recent years there has been heightened interest in the vast selection of peppers — many of which are only available if you grow them yourself. If your only experience with peppers has been green bell peppers for stuffing, roasting, or slicing, you've barely scratched the surface. Peppers, particularly the hot varieties, have gained popularity as more and more ethnic foods come into vogue, including Spanish, Italian, Mexican, Indian, Hunan, Szechuan, Thai, Indonesian, Vietnamese, and Arab dishes. The fruits provide a diversity of shapes, sizes, colors, and flavors that add variety to your garden and pizzazz to your cooking.

A Brief History of Peppers

Peppers are indigenous to tropical America, where they were extensively cultivated centuries before Columbus set foot in the New World. Dried peppers have been found in Peruvian ruins that pre-date the birth of Jesus Christ!

In 1492, when Christopher Columbus and his crew discovered the New World, they also discovered the green and red Capsicums that were common in the West Indies. They assumed from the fruits' pungency that they had found a new variety of the "table pepper" they were so fond of in Europe.

A variety of hot and sweet peppers were found throughout the West Indies, Mexico, Central America, and South America. These quickly caught on with Europeans, who enjoyed them mainly as a seasoning. The fruits soon became synonymous with the ground pepper spice imported from Asia — hence the name "pepper."

Europeans began cultivating their own pepper plants, and carried them to all parts of the globe. Hot peppers were common fare in Africa, the Middle East, India, and much of Asia by the 17th century. In fact, the pepper is one of the New World's major contributions to the cuisine of the Old World.

Although peppers are native to the Americas, it wasn't until after they had grown popular throughout Europe that they were introduced to North America. European settlers brought them back to the New World, and soon peppers were cropping up all over the colonies.

Pepper Classifications

Peppers are members of the Nightshade Family, which includes the tomato, potato, and eggplant. Although they are herbaceous perennials, they are generally grown as annuals.

In parts of Europe, peppers are called *capsicums,* their botanical name. Some historians believe the name came from the Latin *capsa,* or "box," because of their shape. Others believe the name came from the Greek word *capto* — "I bite."

If you thumb through several seed catalogs, you will find that peppers often defy classification — or at least provide some confusion. For example, "Cayenne," "Jalapeño," and "Bell" are all type names, each representing a group of peppers that may contain many cultivars. Yet, a seed catalog may offer a variety called "Cayenne" or "Jalapeño." This list shows common groupings with some representative varieties.

Pepper Group	Representative Varieties
ANAHEIM	Anaheim Chili
ANCHO	Ancho\Poblano
BELL	Big Bertha, California Bell
CAYENNE	Long Red Cayenne, Ring of Fire Cayenne
CHEESE	Yellow Cheese Pimento
CHERRY	Sweet Cherry, Cherrytime
CUBAN	Pepperoncini, Cubanelle
JALAPENO	Early Jalapeno, Jalapeño M
PIMENTO	Pimento Select
SMALL HOT	Serrano Chili
TABASCO	Tabasco
WAX	Sweet Banana, Hungarian Wax

To add to the confusion, the same variety of pepper may have different names in different contexts. For example, an Ancho pepper is the dried version of the fresh Poblano pepper — however, it can be found under both or either name in seed catalogs. This is particularly confusing in ethnic cookbooks. A Mexican cookbook may call for Habaneros; a Jamaican recipe may call for Scotch Bonnets — both are the same pepper.

If you need help with the different groups and names, look for the guide *Capsicum Pepper Varieties and Classification* (Circular #530) posted online at the website of New Mexico State University.

For the sake of simplicity, in this bulletin we'll divide the vast number of varieties into two groups: sweet peppers and hot peppers.

Sweet Peppers

There are probably plenty of green bell peppers in your local supermarket; perhaps some overpriced yellow and red bells, and maybe some Italian frying peppers. Very little else.

If you're looking for green peppers for stuffing, roasting, or slicing for salads or crudités, many bell varieties will suffice. They range in size from the enormous (Big Bertha, King of the North) to the miniature (Jingle Bells, Little Dipper). The green peppers are actually immature fruits. Allowed to ripen, most turn red or yellow with a sweeter, milder flavor and a finer texture. In fact, those expensive "gourmet" red and yellow bells are merely green peppers that have matured.

You aren't limited to the green bells. You can grow purple bell peppers (Purple Beauty, Lilac Belle, Islander), brown (Chocolate Bell, Sweet Chocolate), yellow (Golden Summer, Honey Belle, Orobelle), orange (Ariane, Corona, Valencia), and even creamy white bells (Ivory Charm). Many green bells mature to red early (Ace, Early Thickset, North Star).

Sweet peppers range far beyond the bell peppers. You can select from the long frying types (Cubanelle, Biscayne, Italia), the small cherry varieties (Cherrytime, Sweet Cherry), and the pickling types (Sweet Pickle, Sweet Banana, Pepperoncini).

The following is only a sampling of the many sweet peppers available from seed catalogs.

Variety	Description	Days to Harvest*
ACE	An extra-early bell pepper variety that has high yields even in cold climates—good for Northern gardens. The fruits are up to 4" long with medium-thick walls; ripens to red early. Resistant to blossom drop, so nearly every flower creates a fruit.	50
ARIANE	A Dutch bell pepper variety that ripens into a gorgeous deep orange. The large, blocky fruits are heavy; the flesh is thick, crunchy, and juicy.	68
BELL BOY	An all-purpose bell variety that is prolific, disease and drought resistant. Sweet and mild, medium size, thick-walled, and uniform—perfect for stuffing. All America Winner.	72
BELL TOWER	A midseason bell pepper recommended for all regions. Heavy, thick-walled fruits ripen from dark green to red. Disease resistant.	70
BIG BELL	Noted for its huge yields of sweet bells, ideal for stuffing.	72
BIG BERTHA	The Mother of the Bells! Produces prolific yields of blocky, thick-walled peppers that grow up to 8" long and 4" across—and often weigh over a pound each!	73
BISCAYNE	Cubanelle type, perfect for frying. Elongated fruits are blunt-ended and tapered, usually harvested when light green. Well-branched plants provide canopy to prevent sunscald.	63
CALIFORNIA WONDER	Probably the most familiar and best stuffing peppers. Can be tough to grow in Northern gardens. Fruits are 4" long and 3½" across; excellent cooked or served on a crudité tray	75
CANAPE	Recommended for short-season areas and Northern gardens; tolerates summer heat and drought. Each plant produces about 15 fruits; the peppers are tapered, 3½" long and 2½" wide, ripening to a bright red.	65
CHERRYTIME	An early cherry pepper variety with impressive yields. The 2" round, red fruits are excellent for pickling whole. For sweetest flavor, wait until the fruits are red to harvest.	53
CHOCOLATE BELL	A dark brown bell pepper that's larger and matures later than Sweet Chocolate. The blocky, sweet fruits are 3–4" across. Resistant to mosaic.	70
CORNO DI TORO	The red and yellow "Bull's Horn" peppers from Italy. The 6–8" tapered fruits are twisted like their namesake. Although considered a frying pepper, can be used in salads and (the less twisted fruits) are also great for stuffing.	70
CORONA	Holland import, turns orange when ripe. Medium size, bell-like fruits on compact plants.	66
CRISPY	Early bell hybrid that lives up to its name. The fruits have thick, crunchy walls; turns red early, too. Blocky peppers are 3½" long and 2¾" across.	70

Variety	Description	Days to Harvest*
CUBANELLE	The standard frying pepper. Large, smooth tapered fruits up to 6" long; best harvested at yellow-green stage, when the fruit walls are still firm—they'll be sweeter when they ripen to red, but won't hold up as well.	68
ESPANA	Another frying pepper that produces large yields of tapered, 6–7" fruits with medium-thin walls; ripens from green to red. Excellent disease resistance.	72
GEDEON	Huge, elongated pepper with thick walls; ripens to red.	78
GOLD CREST	Earliest golden bell type. Fruits are blocky and small to medium size. Plants produce heavily and have good disease resistance.	62
GOLDEN BELL	Another golden bell. Larger than Gold Crest with blocky fruits up to 4" across— great for stuffing. Fruits ripen from light green to a deep golden color.	68
GOLDEN SUMMER	Similar to Golden Bell, with large 4"-wide fruits that have thick walls and sweet flavor. Plants grow 20" tall with a canopy of leaves that protects the peppers from sunscald.	70
GYPSY	Wedge-shaped fruits are 4½" long and 2½" wide at the shoulders; crisp, sweet flesh; turns from light green to orange to red. An early variety that can tolerate cool weather and is disease-resistant. Compact plants are great for containers, with as many as 16 peppers maturing at once on each plant. An All America Winner.	65
HONEY BELLE	Large, elongated fruits ripen from green to gold. The plants are vigorous and prolific.	74
ISLANDER	A lavender bell variety with pale yellow flesh; actually ripens in stages from violet to a very dark red. The fruits are medium-sized and have thick, juicy walls.	56
ITALIA	Similar to Corno Di Toro frying peppers and other long, sweet Italian peppers. The fruits are 8" long and 2½" across at the shoulders; ripens early from green to red.	55
IVORY CHARM	A white bell variety that matures from a cream to a soft yellow color with no green at all. The plants produce early and show good disease-resistance.	67
JINGLE BELLS	A miniature bell that is ideal for containers; bears early and prolifically. Fruits are 1½" by 1½" that turn red at maturity.	60
JUPITER	Midseason bell. Produces large, nearly square fruits that are thick-walled and crunchy.	74
KEY LARGO	Cubanelle type, superb for frying. Yellow-green fruits grow up to 7" long and 2½" across at the shoulders, are thin-walled and crisp, and turn bright orange-red at maturity.	66

The table title is: SWEET PEPPER VARIETIES & SOURCES

Sweet Pepper Varieties & Sources		
Variety	Description	Days to Harvest*
KING OF THE NORTH	An early green bell pepper recommended for the North. Fruits are 6" long and about 4" across with thick walls and a mild flavor, particularly after it ripens red.	65
LA BAMBA	Fruits grow up to 5" long and 3" wide with medium-thick walls and a delightful flavor. Great for slicing, stuffing, or roasting.	76
LADY BELL	This bell variety produces abundantly even in short seasons; the fruits are larger and more uniform than most other early types.	71
LILAC BELLE	Another lavender bell that shows no green at all; holds its purple color for a long time before turning a deep red. Excellent disease resistance.	68
LIPSTICK	Elongated, smooth, cone-shaped fruits are 4–5" long, tapering to a blunt point; they turn from a dark green to a glossy red at maturity. The flesh is thick, crunchy, and juicy with a sweet flavor that's perfect fresh for salads or roasted. Produces early; tolerates cool weather. The plants are often covered with the attractive fruits.	53
LITTLE DIPPER	A miniature bell pepper. The fruits are only 2" long and 1I" across, ripening from green to red. Wonderful for stuffing or roasting. Produces early and prolifically.	66
MARCONI	Similar to Corno Di Toro but larger and more flavorful. Up to 12" long and 3" across at the shoulders, ripens to yellow and red. Great for frying or sliced into salads.	70
MONTEGO	A Cubanelle type from the Caribbean. The large, heavy fruits grow up to 9" long and 3" at the shoulders, ripening from white to yellow to red. The thick flesh is juicy and mild, perfect for salads, crudités, or cooking. The compact plants produce early.	60
NORTH STAR	This medium-sized bell pepper is especially recommended for the Northeast; bountiful even in short, cool seasons.	60
OROBELLE	A main season golden bell, slightly larger than other gold varieties. The blocky fruits grow to 4½" long and 4½" across. Grows well in the North.	70
PARK'S EARLY THICKSET	One of the earliest, high-yielding bells. he fruits are medium-sized, maturing to a scarlet red.	45
PARK'S POT HYBRID	This space-saving bell variety is just 10–12" tall, perfect for containers or small gardens. The plants bear heavily and early.	45
PARK'S SWEET BANANA WHOPPER	The first hybrid Sweet Banana—a distinct improvement! The fruits are larger and have thicker walls.	65

	SWEET PEPPER VARIETIES & SOURCES	
Variety	Description	Days to Harvest*
PARK'S WHOPPER IMPROVED	A main season green bell pepper. The fruits are 4" long and 4" across, with thick walls excellent for stuffing or slicing. Matures red.	71
PEPPERONCINI	Traditional, mild pickling pepper recommended for antipasto. Dries well. Fruits are thin and tapered, growing up to 8" long. Picked green for pickling, but will ripen to red.	65
PURPLE BEAUTY	Compact plants have thick foliage to protect the fruits from sun-scald. Peppers are up to 5" long with a rich purple color that holds through cooking. Ripens red.	65
PURPLE BELLE	Another bell variety that changes from green to deep purple to red. The fruits are 3½" long, borne on compact plants.	72
SECRET	Unique color! The fruits have deep purple skin, but the flesh is a light green; ripens to a dark red. Plants produce early and prolifically.	60
SWEET BANANA	Attractive wax-type peppers are great for frying or pickling. Tapered fruits grow up to 6" long with thin walls. Plant may simultaneously bear fruits ranging in color from light green to yellow to orange to red. Can be used ornamentally in the garden.	72
SWEET CHERRY	Like Cherrytime, this variety produces round fruits about 1½" across, starting green and turning red; perfect for pickling.	78
SWEET CHOCOLATE	Similar to Chocolate Bell, only earlier; tolerates cool weather well. The fruits are medium-sized and slightly tapered. The flesh under the chocolate skin is brick red; the peppers won't lose their color when cooked.	58
SWEET PICKLE	Recommended for pickling. Plants bear fruit in a variety of colors at once. Compact plants grow only to 15"—great as an ornamental or potted plant.	65
VALENCIA	An orange bell. The fruits are 4½" long and wide; rich in vitamin A when ripe.	70
VIDI	A French hybrid that, like other European bells, is about twice as long as it is wide.The fruits are 7" long, maturing to red. Good disease resistance; grows well in relatively poor conditions.	75
YELLOW CHEESE PIMENTO	Fruits are large and squash-shaped; ripens green to yellow to orange.	73
YOLO WONDER	A standard green bell pepper that is thick-walled, blocky, growing to about 4½".	76

* Refers to the time after transplanting of seedlings which were started 8–10 weeks earlier. Time to harvest also refers to when the pepper is generally picked; many gardeners may prefer to wait until the pepper fully ripens or matures, which may be weeks later. Check current seed catalogs for availability and pricing. Sources can be found on page 33.

Hot Peppers

Like sweet peppers, hot peppers have many shapes, colors, and flavors. They range in "heat" from the slightly pungent Anaheims to the blistering Habañeros.

Many people believe the smaller the pepper, the hotter it is. Or that red peppers are hotter than green ones. Not true! Serranos are picked green and are longer than the chunky Jalapeños, but they are far hotter than Jalapeños. *Variety* determines how hot a pepper will be, not color, shape, or size.

Growing conditions such as soil quality, moisture, and temperature also influence the hotness of a pepper. A Cayenne will always be hotter than an Anaheim. But a Cayenne grown in New Mexico, where the soil is poor and the climate is arid, will be hotter than the same variety grown in New York, where it is cool and damp. That's why some of the hottest chilis come from Mexico and the Southwest.

What Makes Hot Peppers Sizzle?

Researchers at New Mexico State University have identified at least six compounds in hot peppers that pour on the heat, but the chief chemical ingredient is capsicin, a crystalline alkaloid that acts as an irritant.

The degree of heat in hot peppers can be measured on a scale using Scoville units. Created by William Scoville in 1912, the scale refers to the parts per million of capsicin in a pepper variety. Scoville discovered that the human tongue can detect as little as 1 part per million of the substance. A mild Anaheim may reach 1,400 on the scale, but a Tabasco could reach 50,000 and a fiery Habañero up to 350,000!

HOW HOT IS HOT?

Pepper Variety	Scoville Heat Units
ANAHEIM CHILI	250 – 1,400
JALAPENO	4,000 – 6,000
SERRANO CHILI	7,000 – 25,000
CAYENNE	30,000 – 35,000
CHILE PEQUIN	35,000 – 40,000
TABASCO	30,000 – 50,000
HABANERO	200,000 – 350,000

HOT PEPPER VARIETIES & SOURCES

Variety	Description	Days to Harvest
AJI	*C. baccatum* species; slender, cylindrical, 4" long, ripens to orange-red; medium hot to hot. Cultivated for 4,000 years in the Andes.	80
ANAHEIM CHILI	One of the easiest to grow! Relatively mild; tapered fruits grow to 8" long; matures from green to red. Dry or use fresh. Often used for salsa verde and picante sauces.	80
ANAHEIM TMR 23	Mildly hot Anaheim with medium-thick flesh; fruits grow up to 7" long. Popular choice for commercial canneries.	77
ANCHO\ POBLANO	Best adapted to areas with hot, long summers. Large plants grow up to 4' with 30 or more peppers per plant. Fresh it's called Poblano and is great for chilis rellenos. Dried it is called Ancho and is excellent for moles and sauces. Mildly pungent, heart-shaped fruits are 3–4" long; ripens from blackish green to brownish red.	80
CALIENTE	Productive, early chili that can be used fresh, but is easy to dry and use as seasoning. Straight fruits grow up to 6" long, have thin flesh, and are medium-hot. Ripen to red early.	65
CAYENNE	Pencil-thin, 6" red peppers are smaller than Anaheim and considerably hotter. Excellent for drying and grating for homemade chili powder.	75
CAYENNE LONG SLIM	Another red-hot cayenne variety that is no bigger around than a pencil and matures to a red-orange.	70
CHILE PEQUIN	One of the "bird peppers"—so called because of the fondness birds have for the fruits. Plants grow to 12" tall and 18" wide, bearing a prolific crop of ½" long lava-hot peppers.	80
COPACABANA	Similar to Hungarian Wax, only earlier. The tapered fruits are smooth, up to 8" long, and mature from a waxy yellow to an orange-red.	65
EARLY JALAPENO	Recommended for Northern gardens.	60
FIESTA	An ornamental adapted for container gardening. Can be used in the garden if moved indoors before frost. Compact, bushy plants grow only 9" tall, but produce an abundance of long, slender fruits that change from creamy white to orange and finally to red. Fiery hot!	80
HOLIDAY TIME	Another ornamental (or Christmas) pepper that matures earlier than Fiesta. Green, purple, and red fruits are hot. An All America Winner.	65

HOT PEPPER VARIETIES & SOURCES

Variety	Description	Days to Harvest*
HABANERO	The hottest pepper you can grow! Most famous as the key ingredient in hot Jamaican "jerk" sauces. Belongs to the species *C. chinense;* also called Scotch Bonnets in Jamaica and the West Indies; the major pepper grown throughout Brazil. Needs a long, hot season to fully reach its pungent potential. The small 1–2" fruits are square to bell-shaped, maturing to bright orange, yellow, red, or brown. *Handle with care!*	90
HOT PORTUGAL	Early, produces tapered, 6" long fruits that ripen to a glossy scarlet.	64
HUNGARIAN WAX	A standard, all-purpose hot pepper that can be grown in the North; bares early, produces well in cooler weather, and is prolific. The wrinkled yellow peppers turn deep, dark red at maturity, grow 6–8" long, and are mildly hot. Popular for canning; fruits can be stemmed, seeded, and dried, then ground for homemade paprika.	70
JALAPENO	Perhaps the most familiar hot pepper, particularly in Southwest cuisine. Produces early, setting fiery fruit throughout the summer. Fruits are 2–3" long with rounded tips. Harvested when dark green, but will turn red. Excellent canned or pickled, whole or as a relish; adds distinctive flavor and flame to a variety of dishes and salsas.	72
JALAPENO M	Improved Jalapeño variety. The medium-hot, thick-walled fruits grow 3" long and 1½" wide at the shoulders.	75
LARGE RED CHERRY	Fruits are nearly round, growing up to 1½" wide.	80
LONG RED CAYENNE	Fruits are 5" long but only ½" wide, and are often twisted and curled. Recommended for drying.	75
MANZANO	*C. pubescens* species with unique purple flowers and apple-shaped fruits that ripen to orange, red, or bright yellow. Requires long growing season.	150
NUMEX BIG JIM	Large, slender fruits grow to 10" and ripen to a fire-engine red; mildly hot, recommended for Tex-Mex recipes. Plants produce prolifically.	75
PASILLA BAJIO CHILE	When used fresh, called Chilaca. Cylindrical fruits grow 6–8" long and 1" wide; mild to medium heat. "Pasilla" means "little raisin"— referring to the brown, wrinkled fruits after drying. Delicious fresh or dried; particularly tasty in moles and sauces.	80
PIMENTO SELECT	Mildest pimento you can grow. Fruits are thick-walled, ripening from green to bright scarlet. Recommended for canning.	73
RELLENO	Also known as the "green chili." Long, slender fruits grow up to 6½" long and 2½" wide at the shoulders. Mildly hot; excellent for canning or making chili rellenos.	75

HOT PEPPER VARIETIES & SOURCES

Variety	Description	Days to Harvest*
RING OF FIRE CAYENNE	Earliest and most productive Cayenne variety for Northern gardens. Wonderful, fresh or dried.	60
SERRANO	Similar to Cayenne, only smaller and hotter. The plants are heavy producers of finger-shaped fruits, 1–2" long and slender, usually picked when dark green but will ripen red. Excellent canned or pickled, or thread on a waxed string (dental floss) for drying.	75
SUPER CAYENNE	Plants grow 2' tall—ideal Cayenne for container growing; very attractive when loaded with thin, red peppers. The fruits are 3–4" long and fiery hot. An All America Winner.	70
SUPER CHILI	Hotter than Jalapeños, but not as hot as Cayenne. The cone-shaped, thin-walled fruits are borne upright on the plants, and grow only 2½" long. Harvest when green or red. As a bonus, the plants work well ornamentally, even indoors in containers.	62
TABASCO	*C. frutescens* species. Infamous as the main ingredient in the fiery sauce that shares its name. The small, pointed fruits mature from yellow to orange to red. The plants grow up to 3' tall; each plant can produce over 100 peppers. The fruits are easiest to dry right on the plants. Produces best where the season is long and hot.	80
TAM MILD JALAPENO (JALAPA)	A milder variety for those who like the flavor of Jalapeño but not the heat. Heavy yields of 2½" fruits; can be used when green, but better flavor when red. Excellent when used on pizza or nachos, in sauces or pickles, or simply sliced with Monterey Jack cheese.	70
THAI HOT	Almost as hot as Habaneros! The 1" fruits are like firecrackers! The mound-shaped plants reach a height of 18", blanketed with tiny peppers of green and red held erect above the foliage. The plants can be used ornamentally, particularly in patio containers or even in an annual flower bed. Bring in the house before the first fall frost to enjoy as a houseplant. Widely used in Oriental dishes; but go easy—a few fruits go a long way.	75
YELLOW CAYENNE	Tapered fruits are 8" long and ¾" wide; ripens from yellow to dark red.	68
ZIPPY	Similar to Cayenne, but much milder. Fruits are 6" long but only ⅝" wide. Usually harvested when green, but matures to red. More zesty than hot, the peppers can even be used in salads or crudité trays.	57

* This refers to the time after transplanting of seedlings which were started 8–10 weeks earlier. Time to harvest also refers to when the pepper is generally picked; many gardeners may prefer to wait until the pepper fully ripens or matures, which may be weeks later. Check current seed catalogs for availability and pricing. Sources can be found on page 33.

Planning for the Best Production

With peppers more than most other vegetables, matching the right variety with the right location is essential. But, before you open a seed catalog, first consider *your* needs.

The kitchen is the best place to plan your pepper crop. How do intend to use the peppers you grow? Do you want to grow sweet peppers, hot peppers, or both? Do you want to stuff, roast, or fry them? Use them in hot sauces? Slice them for fresh salads? How will you store them — frozen, pickled, or dried? Do you want to dry them for decorations?

Consider also your garden space. How many plants can you grow? Do you want to use peppers ornamentally in your flower beds? Will you grow them in containers or indoors?

If you're satisfied with green bell peppers, a good nursery will have some varieties. But if you want a greater variety, you'll have to start from seed. Here is where planning is key.

Peppers vary widely in their regional adaptation. California Wonder may be wondrous in California, but it has lackluster production in Maine. If you live in a short-season area, in the North or in the mountains, select early varieties that mature in 50–70 days. If you live in the South or in a long-season area, choose main season and late-maturing varieties.

For advice on what varieties are best adapted to your local environment and soil conditions, talk with fellow gardeners, your County Extension agent, or your state university Extension Service. Look for varieties that are resistant to common diseases, notably Tobacco Mosaic Virus or Potato Virus Y.

I like to experiment, searching by trial and error for the best varieties for my garden. Grow a few different varieties each year and you'll quickly discover that some produce more or taste better than others. Eventually you'll find some "favorites" that you can count on year after year!

When ordering pepper seeds from catalogs, keep in mind:
- A single seed packet contains more seeds than most home gardeners can use in one season. Pepper seeds can be saved for up to four years, but they become increasingly less viable. Buy one or two fresh packets each year and trade different varieties with friends.

- Unless the seed catalog states otherwise, the "number of days to maturity" refers to the days after transplanting until the plant produces a full-sized fruit. You must add 8–10 weeks (50–70 days) for the time between sowing and transplanting, and 3–6 weeks (20–40 days) if you want the fruits to fully ripen.

Getting a Jump on the Season — Starting Peppers Indoors

In areas that don't have long, warm seasons, peppers must be started indoors to produce a harvest before the fall frosts.

Many people think peppers should be started at the same time and under the same conditions as tomatoes. But peppers take longer and require a higher temperature to germinate and are more finicky concerning climate. You've got to start them earlier and transplant them later than tomatoes.

Establish the last expected frost date in your area (your Extension agent or fellow gardeners can be of assistance), then count backwards 8–10 weeks to determine when to start your pepper plants. For many of us, that means January or February. If you receive your seed catalogs in November or December, you must act quickly.

Seeds will sprout in about a week at a temperature of 70°–80°F. Keep in mind that sprouting is uneven at any temperature, and that germination rates vary from variety to variety. Hot peppers can be *very* slow to germinate.

You can start your seeds in flats, but you'll have to transplant them into pots when their first true leaves appear. I prefer to soak the seeds, pre-sprout them, then plant them in individual containers.

To break the seeds' dormancy, soak them for a few hours in lukewarm water. For the slow-to-germinate hot pepper varieties, soak them in warm, salted water (1 tbsp. per quart) for 2–3 days. (Don't leave the seeds in the solution for more than 3 days or they may not germinate at all!) After soaking, place the seeds between moist sheets of cloth or paper towel, put them in a plastic bag, and put the bag in a warm place. The top of a refrigerator is perfect.

Check the seeds frequently. When they sprout, plant them ¼-inch deep in peat pots, six-packs, or other small containers filled with a commercial potting soil or an equal mix of friable loam, leaf mold, and sharp sand. As with any seedlings started indoors, you must guard against damping-off disease, a soil-borne fungi that can decimate your crop. The best precaution is to use a sterile potting mix and be careful not to overwater.

Peppers are fussy about temperature in the seedling stage. After you've planted your sprouts, provide bottom heat at about 68°F, if possible. Once the true leaves appear on the seedlings, move them to a location that receives full sunlight — a sunny southern window with no cold drafts will suffice. Daytime temperatures should average 70°–80° and night temperatures about 60°–70°F. Lower temperatures may inhibit growth and higher temperatures may make the plants weak and scrawny at transplanting time; both will reduce yields later in the summer. Water the seedlings from the bottom and fertilize lightly until you can move the plants into the garden.

Preparing the Perfect Garden Site

Choose an area where the peppers will receive a maximum amount of direct sunlight. For best production, peppers need 6–8 hours of full sun each day — much the same as tomatoes.

Although peppers often grow well in poor soil (particularly the small, hot varieties), they produce higher yields when provided the proper soil conditions. Peppers are heavy feeders that produce a long harvest, so don't skimp on soil preparation.

Peppers prefer a deep, fertile (but not too rich), sandy or gravelly loam that drains well. If your soil is heavy, a raised bed is your best solution — fill it with a mixture of loam and sand. If you can't construct a raised bed, be sure to incorporate massive amounts of organic matter into the bed, preferably during the preceding fall, to give it plenty of time to work its magic. If your soil is excessively sandy, the organic-matter treatment will provide needed nutrients and help the soil retain moisture.

Whatever your soil conditions, you have everything to gain by preparing the soil with organic matter. Peppers need plenty of nutrients to reach optimum size and production, and they need it steadily throughout the season. If the soil is too rich with nitrogen early in the season (as can happen with chemical fertilizers) the plants may produce luscious foliage at the expense of flower and fruit production. This isn't as likely to happen with a balanced mix of organic matter.

If you have disease problems in your garden, refrain from using compost in your pepper bed, particularly if you've dumped potato peelings in the compost pile. Even the hottest compost can't maintain high enough temperatures at the edges of the pile to eliminate all bacterial diseases. If you want to add organic matter, instead of compost, use well-rotted manure, well-rotted sawdust, or leaf mold.

I've found a combination of organic and slow-release chemical fertilizers works best. When the temperatures stabilize in the early spring and I can work the soil, I blanket the area with about 2 inches of compost, sprinkle it with 5-10-10 fertilizer, and work it all in. If weather permits, I do this about a week before transplanting, as the peppers are hardening off.

Peppers are moderately tolerant to an acid soil (pH 5.5–6.8), but highly acidic soils should be limed according to soil test recommendations. Dolomite limestone contains the magnesium that peppers need to prevent leaf drop, which results in sunscalded fruits.

In the North, for early peppers in cold soils, cover the prepared bed with a plastic mulch at least a week before transplanting your seedlings. This will heat the soil nicely and will be less of a shock to young pepper roots. A brown polyethylene mulch is preferred to black or clear plastic mulches; it doesn't build up as high a soil temperature and has been shown to increase yields by 20 percent.

Moving Seedlings Into the Garden

Just like moving your family into a new home in a new neighborhood, moving your peppers into the garden can be traumatic at first. Make the move as stress-free as possible.

Timing and Temps

When I first started growing peppers, I made the mistake of transplanting the peppers at the same time as the tomatoes. Although I often had tomatoes in June, I also had quite a few pepper crop failures, particularly after a cool, rainy spring. I soon learned that, with peppers, it's smarter to wait until the weather is warm — and will stay warm!

Peppers will never fully recover from a cold shock. If they're not killed outright by an unexpected frost, they'll produce poorly all summer. The plants grow best with daytime temperatures of 70°–80°F, so wait until well after the last expected frost in your area before transplanting.

Before transplanting, harden off your pepper plants to get them acclimated to the garden. Place seedlings in a warm, sheltered area that is partially shaded — they should receive sun for a few hours each day. Move the plants back indoors overnight if temperatures are expected to drop below 60°F. After a week or so of this hardening-off process, the pepper plants should be accustomed to outdoor conditions.

When the weather is consistently moderate, you can move the plants into the garden. I usually wait until night temperatures are above 60° and never fall below 55°F. A soil thermometer is helpful. When the soil 4 inches below the surface is 65° or higher in the morning, you can transplant your peppers. If you've prepared raised beds covered with a plastic mulch, you may not have to wait as long!

Transplanting and Spacing

The ideal pepper transplant has about 5 true leaves, is as tall as it is wide, and has only a few tiny flower buds or none at all. Plants sold at garden centers are often large and loaded with blossoms. The root systems are too small to support the plants in the garden. You probably can salvage such a plant by trimming back a substantial amount of the foliage before transplanting. This gives the roots time to spread through the soil, and encourages a bushier growth and better production.

Studies at the University of Texas and the University of Florida show that close spacing results in more and larger fruit. The leaves of each full-grown pepper plant should be touching those of surrounding neighbors. I've found that most bell peppers should be spaced 18 inches apart and hot peppers about 12 inches.

Peppers need a healthy dose of nutrients at transplanting time to put on enough growth to support large yields. For each plant, dig a hole about 6 inches deep. Add a 2-inch layer of organic matter and a handful of 5-10-10 fertilizer; mix it well with the soil at the bottom of the hole. An old gardening trick is to toss in a book of matches

— the sulfur in the matches will make the soil more acidic, which will please the pepper. (Make sure there is some soil between the matches and the roots of the transplant.) A sprinkling of colloidal phosphate will help prevent blossom end rot.

Set each seedling lower in the ground than it was in its pot. If your pepper plants are in peat pots, be sure to bury the entire pot below the soil surface. If any of the peat pot material is above the soil surface, it will act as a wick, drawing water from the plant and eventually killing it. Backfill around the plant and carefully tamp it down with the heels of your hands. Mulch lightly around the base of the plant.

After Transplanting

Immediately after transplanting, water thoroughly to remove any air pockets in the soil and help settle in the roots. The plant will quickly develop a sturdy root system and will soon tap the nutrients you placed at the bottom of the hole.

To protect young plants, you may want to cover the bed with a floating row cover like Reemay to deter flying insects, keep the plants warm, and prevent wind damage. When the peppers begin to blossom, remove the cover to allow bees and other insects to pollinate the plants.

If an unseasonably cool night threatens after you've transplanted the peppers, protect the plants with Wall O' Waters, cloches, plastic milk jugs, or another form of heavy covering. Don't forget to uncover the plants the next morning or the plants may cook in the noonday sun!

What Do Peppers Need?

Water!

Remember, peppers are natives of the American tropics, where the humidity is high and it rains almost daily from May to October. Peppers need water from the time you transplant your seedlings until the end of the season. How much depends on where you live and what kind of summer you're having. But wherever you garden, the key to watering peppers is *moderation*. If the soil is too dry, the plants will wilt and refuse to produce fruit. On the other hand, peppers won't tolerate waterlogged roots. The plants will start shedding leaves, exposing fruit to sunscald, or will simply shut down altogether. Too frequent waterings also leach water-soluble nutrients from the topsoil.

A deep, weekly watering that is the equivalent of 1 inch of rainfall is preferred. However, if it is unusually hot and dry for long periods, or if your garden soil is sandy and drains too quickly, you may need to water more frequently. Peppers respond splendidly to trickle or drip irrigation.

The critical time for watering is from flowering through harvest. If the plants are stressed from lack of water, many buds and blossoms will drop. Don't heed the common belief that you can ripen fruits sooner by withholding water or otherwise stressing your pepper plants. Pampered plants will bear far more and better quality peppers.

The Importance of Mulch

Watering may not be enough during the hot, dry, breezy days of summer when plants transpire moisture rapidly from their leaves. In addition to water in the soil, there must also be enough humidity — the primary benefit of mulch.

A thick, finely-shredded, organic mulch not only keeps weeds at bay and helps the soil retain moisture, it also keeps the humidity high around your peppers. The idea is to supply a wide evaporative area under each plant that will continually pour on the humidity. A mulch made of a mixture of grass clippings, straw, and shredded organic matter works best in my garden. Starting an inch or two from the plant's stem, apply the mulch about 6 inches deep, extending just past the dripline of the plant.

Wait until summer heat begins to peak before applying such a mulch, however. A thick organic mulch will keep the soil cool around the roots — great during the summer, but not in the spring when you want the soil warm. Before applying organic mulch, remove any plastic mulch. This allows the organic mulch to add nutrients to the soil beneath as it breaks down.

Providing a Balanced Diet

Like most vegetables, peppers need *nitrogen* for sturdy stems and foliage, *potassium* for strong roots, and *phosphorus* for fruit production. The trick is to time feedings right.

Give the plants potassium and phosphorus initially and later in the summer when blossoms appear, with small doses of nitrogen. The 5-10-10 you applied at transplanting time will get the peppers off to a good start. In the third month of growth, when blossoming starts, the plants need more nutrients. Pull back the mulch and side-dress the plants with a sprinkling of 5-10-10 and some bone-meal around the dripline — don't fertilize at the base of the plants. Carefully work the mixture into the soil and replace the mulch. When the fruits are about an inch long, repeat the process.

Peppers often produce lots of flowers but few fruits. This is generally caused by a lack of magnesium. To jump-start production, spray your plants with a solution of Epsom salts just as they start to blossom. This topical application will provide all the magnesium required. Mix 2 teaspoons of Epsom salts in a quart of warm water and spray it on the leaves and blossoms; repeat the process two weeks later. The plants will turn a dark green — followed by a flush of fruit!

Other Growing Techniques

- Large pepper plants laden with fruit should be staked or caged for support. Peppers are among the most brittle of vegetable plants and are easily damaged when bearing heavy yields, handled roughly, or even brushed against. When tying a plant to a stake, use strips of old nylons which can expand as the stems enlarge. Don't use wire or string which will gradually choke off the plant.
- When planning your garden, remember to include peppers in your succession plantings. Because peppers shouldn't be transplanted until the weather is consistently warm (in May or June for many of us), they can follow fast-growing, cool-weather crops like spinach, lettuce, and radishes.
- Temperatures over 90°F can cause buds and blossoms to drop, particularly if the air is dry. If this is a common problem in your garden, plan to offer your peppers some shade

during blazing afternoon sun. Plant them on the north side of tall plants like sweet corn or a trellis of cukes or pole beans. The peppers should be shaded from about noon to 3 p.m., when the heat is most intense. Don't completely shade the peppers, however; they should receive full sunlight as the sun moves west later in the day.

- Many gardeners claim that companion planting will enhance pepper production. Try planting peppers near carrots, basil, parsley, and tomatoes. Peppers share space especially well with onions and cabbage. However, keep the plants away from kohlrabi and fennel — traditional pepper "enemies."

Using Peppers Ornamentally

Peppers blossom and fruit simultaneously — and usually with abundance. The white flowers and colorful fruit are just as welcome in the flower garden as they are in the vegetable garden. Look to the sunny "holes" in your landscape that can be strategically filled with lush pepper plants.

Peppers are bushy, dark-green plants with upright growth. They can be used much like shrubs. The fruits come in all sizes, shapes, and colors, and are often borne above the foliage for an attractive display — even the green fruits are showy. Many varieties with compact growth and colorful fruit (Corona, Gypsy, Purple Beauty) work well in a flower border. I particularly like the varieties with plants that grow no larger than 18 inches, like Sweet Pickle, Thai Hot, Chile Pequin, and Super Chili. They require little space and produce hundreds of small, cheery peppers that add a novel accent to any ornamental border. Try combining them with dwarf zinnias, marigolds, portulaca, red verbena, or lobelia for a splash of vibrant color. And while you're enjoying the display, don't forget to harvest a few of the peppers to enjoy in the kitchen.

Growing in Containers — Indoors and Out!

Many peppers are adaptable to containers, pots, planters, and window boxes. Don't forget to bring one or two indoors before the end of the season to extend fruit production.

Whether indoors or out, peppers need plenty of space for root formation. I plant mine in pots at least 12 inches deep. Put some pot shards or pebbles in the bottom of the container for drainage, then fill the container with a good potting mixture of 1 part well-rotted compost, 1 part coarse sand, and 1 part soil — a commercial soil mix will work fine.

Outdoors

Peppers grown outdoors in containers can save limited garden space. They have the extra advantage of portability, particularly if the pots are placed in a small wagon. This permits you to move the plants to ideal locations throughout the season — into full sun on a cool spring day, shade during blistering August afternoons, or into the flower garden where their bright, colorful fruits can be used ornamentally.

Peppers in containers make excellent patio/deck plants. A large pot can accommodate a small collection of peppers and some basil or other herbs. I've enjoyed Gypsy and Cayenne plants in containers with annual flowers that cascade over the rim, like alyssum, lobelia, and browallia. A single stake for the central pepper plant will keep it erect even under a heavy fruit set.

The small peppers grow well in window boxes with petunias and miniature zinnias. Select varieties that bear abundant, small fruit above the foliage, such as Super Chili, Thai Hot, and Sweet Pickle.

Indoors

Although I have grown some peppers from seed as houseplants, I prefer to move a few garden plants indoors before the end of the season. Because peppers are perennials, you can even replant them next year in the garden.

Bring the plants indoors while the weather is still mild, weeks before the first expected frost in your area. The best time to dig up the peppers is on a cloudy day or near dusk. Soak the soil around the peppers about two hours before digging to lessen the shock to the roots. Work the spade around the dripline of each plant, digging straight down, and then carefully lift the plant out of the ground.

Before potting, prune the old anchor roots, but don't touch the feeder roots near the soil surface. Also prune back the foliage until the plant is about the size of the pruned root ball. Place the plant in the container and pack the potting mixture around the root ball. Then fill the container to within an inch of the rim, soak the soil until water drips from the drainage holes, and gradually bring it indoors. You need to give the plants time to adjust — like hardening-off in reverse.

Bringing peppers indoors may delay fruiting until the plants are fully acclimated, but with pampering they will eventually produce fruit while providing attractive houseplants. The fruits may not be as large as they would be in the garden, but if you notice diminishing fruit size, the plant may need more nutrients, steady moisture, or sunlight.

Use a slow-release fertilizer high in potassium and phosphorus (5-10-10) to promote flower and fruit production. Or use a liquid fertilizer in a weak solution with every watering. Be careful not to overwater — a deep soaking once a week is generally sufficient.

One or two plants indoors will probably be enough, particularly since peppers require plenty of direct sunlight and will hog window space. Supplemental lighting helps, but full sun will help the plants set more fruit. The plants should ideally have 3,000 foot-candles of lighting for flowers and buds to form. Less than 2,000 foot-candles will keep the plants green and bushy, but fruitless.

Peppers require an optimum temperature of 75°–85°F and constant humidity for the best production. During the winter, our homes may be too dry and cool. Keep the plants away from extreme temperatures — don't place them near drafty windows or next to a hot air duct. The best way to increase humidity around the plants is to place the pots in trays filled with pebbles, and keep the pebbles moist to allow evaporation. Don't keep the pots in standing water, however.

In the garden, bees and other insects pollinate the pepper blossoms. If you want your indoor peppers to fruit, you'll have to hand-pollinate the flowers. Simply use a cotton swab or a soft paintbrush to transfer pollen from one flower to another. Peppers are self-pollinators (they don't need pollen from another pepper to fruit), but they still need some help.

By growing peppers in containers indoors, Northern gardeners can try to grow some of the long-season hot peppers. And for those with little or no garden space (like many apartment or condo dwellers), indoor peppers are a big plus!

Pepper Pests & Diseases

Although they are generally pest- and disease-free, peppers can be affected by the same insects and diseases as tomatoes. Most Northern gardeners have few problems with pepper pests, but Southern gardeners are often not so lucky.

Bad Bugs!

Insects that attack peppers include cutworms (early in the season), aphids, pepper maggots, pepper weevils, tomato horn-worms, Colorado potato beetles, leaf miners, flea beetles, and corn borers. Nearly all of these pests can be controlled using organic insecticides like Bt (Thuricide) for the caterpillars and rotenone or pyrethrum for most of the others.

Another option is to spray the foliage with your own hot-pepper concoction. Combine a handful of your hottest chili peppers, several cloves of garlic, a tablespoon of non-detergent soap, two or three tablespoons of pyrethrum powder, and three cups of luke-warm water. Liquefy in a blender, put through a strainer, and then spray on the pepper leaves, tops and bottoms, every few days. This homemade pesticide will keep most pepper pests at bay. (Use an old blender that you no longer use in the kitchen.)

For heavy infestations, you may have to resort to stronger chemical pesticides. Use them according to the manufacturer's instructions, but be careful: Many commercial products, including the popular Sevin, are harmful to bees and other beneficial insects. Contact your County Extension agent concerning the latest insecticides that are the safest and most effective for your home garden.

One way to keep your peppers pest-free is to relentlessly weed your garden, removing hiding and breeding areas for bugs. As a precautionary measure, till under or burn all crop residues in the fall. This destroys many insects that would winter-over in the garden to create havoc next year.

Common Pepper Diseases

Peppers also share some diseases with tomatoes, eggplants, and potatoes, such as anthracnose, bacterial spot, blossom-end rot, and mosaic. There are four easy ways to keep disease problems under control:

- Plant disease-resistant varieties.
- Rotate your crops on a three- or four-year cycle. Try not to plant any member of the same plant family in the same bed more than once every three years. Many diseases are soil-borne and plant-selective. Crop rotation greatly reduces the occurrences of disease.
- Weed the garden. Weeds can spread fungi and viruses (and harbor insects that carry diseases) to nearby healthy pepper plants.
- Avoid working in the garden when it's wet. Diseases spread more rapidly among wet leaves.

Harvesting Peppers

Peppers are fruits, like cucumbers, that are traditionally harvested in an immature stage. Many gardeners, however, have learned that allowing peppers to ripen fully on the vine improves their quality. Peppers can be harvested any time, but the flavor doesn't really develop until maturity.

The dilemma is this: Prompt harvesting improves yields. If you allow the fruits to mature on the plants, they will produce fewer peppers. If you pick the fruits throughout the season, the plants will continue to produce — but you may have fewer ripened peppers. The determinant may hinge on where you live.

In mild, long-season areas, you can allow the peppers to ripen fully before harvest and still expect a second crop. In short-season areas, however, yields will be reduced if you leave all the fruit on the vine until full maturity. The plants wouldn't have time to flower and fruit again before a killing frost. You may have to sacrifice high yields for taste — or vice versa. Seek a balance; leave some peppers on each plant to ripen, and harvest the others throughout the season as they become table-sized.

When picking peppers, it's easy to break off a branch or uproot the whole plant if you're not careful. Use a sharp knife or heavy scissors or shears to harvest the fruits, cutting the tough stems rather than tugging on the plants and risking damage. Leave about ½-inch of stem on each pepper. When frost threatens, harvest the remaining fruit — the plants won't survive freezing temperatures. If possible, eat the peppers on the day you pick them. If not, leave them on the kitchen counter where they'll continue to ripen for the next few days. Don't enclose them in plastic wrap or a plastic bag or toss them into the vegetable crisper of your refrigerator! If you have too many peppers to use right away, consider the following storage options.

Storage

Freezing

Probably the easiest way to store peppers is in your freezer. Although peppers can be frozen whole, they hog freezer space. A little preparation is recommended.

Choose firm, thick-walled fruits that are blemish-free. Wash them, cut out the stems, remove the seeds, and cut them according to intended use. If you want to use them for stuffing, cut them in half lengthwise. If you want to use the peppers for seasoning in winter soups, stews, casseroles, and spicy sauces, dice them. Cut them into rings or strips for frying. You don't need to blanch the peppers; just pack them into freezer containers or bags; seal, label, and freeze.

Frozen peppers will be soft when thawed, but the flavor will still be there.

Canning

You can also preserve peppers by canning them. Because they're low-acid fruits, they must be canned under pressure, however. I find it easiest to pickle them. You can pickle your peck of peppers in much the same way you would cukes — either in a crock or processed in jars.

Select a large crock with a lid. Wash the container thoroughly with salt water to make sure it is sterile. Wash the peppers as well, and preferably leave them whole as you put them in the crock. If you're using small peppers, cut two slits in each fruit to allow for complete pickling. If the peppers are extremely large, stem and core them, then cut into quarters. I like to add a few small hot peppers to a crock full of sweets, just to spice things up a bit.

For a simple brine, mix 4 cups of water, 4 cups of vinegar, and ½ cup of pickling salt. Add a clove or two of crushed garlic and some fresh herbs. Many people use dill as they would with cukes. But for a really different, pleasant flavor, try adding a few sprigs of fresh mint. Pour the brine over the peppers and cover. Place the crock in a cool place out of the sun and the peppers should be ready to eat in about 2 weeks.

You can also store your pickled peppers in jars. Simply fill cleaned jars with your peppers, then add the garlic and herbs. Bring the brine to a boil and pour it over the peppers, covering by ½-inch. Screw on the caps and process in boiling water for 15 minutes. Store the jars in a cool place.

Drying

Peppers, particularly the thin-walled hot varieties, are easily sun-dried or hot air-dried. It's best to use fully-ripe, red or orange fruits. Even sweet peppers can be dried, but I've found that hot peppers work best.

The easiest way is to allow the fruits to dry right on the plant — either in the garden or by harvesting the entire plant and hanging it upside down in a place that's warm and dry with good air circulation. An open garage or garden shed will work nicely in late summer. Many gardeners thread peppers on a string (dental floss is recommended) and hang them in the kitchen to dry while using them as decoration.

When using tough-skinned peppers, like most of the bell varieties and some hot peppers like Jalapeño and Poblano, it's best to peel them before drying. If you roast them first, the skins are easily removed and the peppers retain that wonderful roasted flavor as they are dried.

Most small peppers can be dried whole, but the larger types need some preparation to speed up the process. Select fully ripe, unblemished fruits. Core and cut the peppers into rings or strips. Then spread them evenly in a single layer, with none of them touching, on a rack to dry. This can be done in the oven at a low temperature for 6–12 hours, in a food dehydrator, or even outdoors in a cool, dry, airy spot that receives direct sunlight. Turn the peppers occasionally to make sure they dry evenly.

The key to drying peppers is to do it *s-l-o-w-l-y*. The slower they dry, the better they'll retain their color and flavor, and the longer they can be stored. When the peppers are crisp and brittle, they're ready for storage. Place them in airtight containers and store them in a cool, dark area.

When you want to use your dried peppers, you have many options. You can use dried strips as seasoning on meats. Soak whole dried peppers for a few hours (or simmer them) until tender to be used for stuffing. An excellent way to use dried peppers is for grinding into your own powdered seasonings. Remove the stems and seeds from the peppers and grind them to the preferred consistency. A small electric coffee grinder will work fine, but be sure to clean it thoroughly after use. If you're grinding hot peppers like Cayenne, be sure to wear rubber gloves during the process. And be careful not to inhale the powder, or you'll be in for a very nasty surprise! Store the ground peppers in airtight containers to preserve their flavors and pungency.

How Peppers Benefit Your Health

Peppers have the highest vitamin C content per pound of any fruit or vegetable, including citrus. A pound of red peppers offers almost three times the amount of vitamin C as a pound of tomatoes. Peppers also provide vitamin A, niacin, riboflavin, thiamine, a variety of minerals, and a good dose of fiber.

The longer a pepper matures on the plant, the greater its vitamin content will be. The vitamin C in a red bell pepper is almost double that of a green one of the same variety. Just ¼ cup of diced, raw red pepper contains about 150 percent of the RDA for vitamin C and 100 percent of the RDA for vitamin A.

The capsicin found in peppers (particularly the hot ones) is currently being studied for its health benefits. We know that the alkaloid has positive effects on the digestive system and, despite popular belief, there is no scientific evidence that peppers irritate ulcers.

Hot peppers may also be good for your heart. Studies suggest that capsicin reduces clotting and lowers fat levels in the blood. There is also evidence that capsicin has antibacterial properties and may play a role in cancer prevention.

Dieters take note — a whole green sweet pepper contains only 15 calories, raw or cooked! Remember, however, too many peppers, especially green ones, can cause stomach distress. And with hot peppers, a little goes a long, long way.

Kitchen Preparation: Where Peppers Shine!

It's not the pepper's nutritional value that entices gardeners, however — it's the versatility of peppers in the kitchen!

To prepare bell and other large peppers for cooking, first wash them thoroughly and core them, removing the seeds and the white inner ribs, leaving only the shells. Pepper seeds are extremely bitter, and seeds from some hot peppers are too hot to eat. Best to discard them.

Be cautious when handling hot peppers. For the real scorchers like Cayennes and Habañeros, use latex or plastic gloves. The capsicin in the peppers can actually "burn" your fingers, sometimes for days afterward. The volatile oils can be transferred from your hands, so don't touch any part of your body, especially your eyes. Cleaning hot peppers under cold running water may keep your hands cool, but the capsicin is not water-soluble, so you should still be careful. If you find your hands are burning, try washing them with something acidic like vinegar or lemon juice to remove the oil.

Likewise, be cautious when *eating* hot peppers, particularly with varieties you've never tried before. If you taste a pepper that is far too hot for your tolerance level, *don't try to cool it with a drink of cold water!* Water will only spread the capsicin around your mouth. You should immediately eat an oil-absorbing food like potatoes, pasta, or bananas, but the best antidote I've found is a spoonful of chilled, plain yogurt.

Stuffed Peppers

The thick-walled bell peppers are perfectly shaped for stuffing. The yellow, red, and orange varieties add color to the meal. Once the peppers are cored, select fruits that will stand up without tipping over, or cut the fruit in half lengthwise to form a "boat."

Blanch them in boiling water for about five minutes, then plunge them in ice water to halt the cooking process. Drain and pat the fruits dry before stuffing, either with the traditional meat and rice or any combination of cooked vegetables, such as succotash or mashed potatoes and cheese. Bake in a preheated oven at 375°F for about 20 minutes or until the peppers are tender and the stuffing is heated through.

Broiled and Roasted Peppers

The best way to experience the full-bodied, robust flavors of peppers is to broil them in the oven or roast them on a grill.

When roasting peppers on an outdoor grill, use wood charcoal, if available, instead of briquettes. Avoid using lighter fluid, which often lends an unpleasant taste. Lay the peppers on the grill close to the glowing coals, turn them with tongs until charred on all sides, and pop them into a paper bag for about 10 minutes to allow the steam to loosen the skins.

To broil peppers in the oven, cover the broiler pan with a sheet of aluminum foil, arrange the peppers on it, and place the pan about 3 inches from the heating element. Turn the peppers to char them on all sides. Then remove from the broiler. Crimp the foil around the peppers, enclosing them, and leave for about 10 minutes to steam the skins loose.

Once broiled or roasted, the peppers are ready to be cored and peeled. Remember to wear rubber gloves if you're handling hot peppers. Cut around the stem of each pepper and pull out the seeds; scrape out the inner white membranes. Peel the skins either by using a knife or sliding them off under cold, running water. You can use the peppers whole or slice them into strips to use in a variety of dishes — try tossing the strips in olive oil, with a pinch of salt, and some crushed garlic. Roasted peppers can also be stored in the freezer in plastic bags for six to eight months.

Use Your Imagination!

Peppers can be used in hundreds of ways, so don't be afraid to experiment with a variety of dishes.

To create a hot sauce that works well with many oriental dishes, steep some small hot peppers in vinegar. Pack a small bottle with Thai Hot, Chile Pequin, or other tiny, hot peppers. Fill the bottle with vinegar, cork it, and let it steep for a few weeks in sunlight. The resulting liquid should be used sparingly, it is so hot! You can refill the bottle once or twice more with vinegar to create new batches. A splash of this in your next stir fry will have more than your wok heating up!

Add dried peppers to meat dishes such as chicken, steak, and even fish, or to cooking sauces, marinades, and dressings.

Fresh peppers, sweet or hot, are the best. They can be used whole, quartered, sliced in strips or rounds, or chopped. They can be tossed in salads, pickled, sautéed lightly in oil with a touch of garlic, added to zesty sauces, stuffed with a combination of meat and rice and baked, roasted and skinned, used to top pizza, sliced thinly to add to sandwiches, used to liven up an omelet, or offered raw accompanied by a dip for appetizers. The list goes on and on.

No matter how they're used, peppers really punch up the menu.

Sources

Fedco Seeds
207-426-9900
www.fedcoseeds.com

Johnny's Selected Seeds
877-564-6697
www.johnnyseeds.com

Pepper Joe's
888-660-2276
www.pepperjoe.com

Reimer Seeds
mail@reimerseeds.com
www.reimerseeds.com

Tomato Growers Supply Company
239-768-1119
www.tomatogrowers.com

Veseys
800-363-7333
www.veseys.com